Today is the Perfect Day to Improve Customer Experiences!

Understanding how customer experiences go wrong, so yours can go right!

by Lisa D. Dance

Illustrations by Antonio Meza

TABLE OF CONTENTS AT A GLANCE

Content Warning: The stories in this book illustrate the frustration and powerlessness customers and employees sometimes feel. If reading them becomes stressful, please step away and take time to destress.

TABLE OF CONTENTS

Content Warning: The stories in this book illustrate the frustration and powerlessness customers and employees sometimes feel. If reading them becomes stressful, please step away and take time to destress.

TABLE OF FIGURES

INTRODUCTION

I've been studying customer transactions since I was nine years old, when I told my mother I wouldn't shop at a particular store anymore because of how they treated their customers. I noticed how difficult the store made it for customers to get a refund, even when they followed the refund policy.

Even at nine years old, I knew that was not how you treated a customer. It didn't seem like a smart idea either. Why only give refunds begrudgingly when any other day these customers were making purchases? Why make the experience so unpleasant? Customers wouldn't just stop making returns. They would stop making purchases, too. This started me on a path to not only study customer interactions, but also to try to figure out ways to improve them for customers and the business. Because even at nine years old, I knew both customers and the business needed to achieve what they wanted.

Figure 0.1 Photo of Lisa D. Dance, about 9 years old

Eventually, this interest led me to work in the User Experience field, where my goal is still to understand customer wants and needs and business goals, and to craft solutions that meet them. While the User Experience field is typically thought of as relating only to online experiences or technology, in our increasingly interconnected world, my work bridges the online and offline experiences.

Like the stories in this book, my work also delves into areas considered to be Customer Experience (all the interactions a customer has with a company) and Service Design (designing services that consider the organization's people, processes, and systems as well as customers' experiences).

So when I started to notice as a customer, that customer experiences seemed to be getitng worse not with just one company but many, I was continuing the pattern of observation I started when I was nine. I started talking with friends and family, and everyone seemed to have a story or two—or nine—about a customer experience gone wrong and how much work they had to do to get issues resolved. I began collecting these stories and tried to understand what was happening and why. Some patterns began to emerge around what was going wrong.

I eventually started sharing what I was seeing in social media posts, then in an article, and a presentation, but there was still a "bigger yet more nuanced idea" that needed to get out about how technology plays a part in making the customer experience more difficult.

Then I met Antonio Meza, a talented Business Consultant with Visual Superpowers. A call led to a collaboration where we used empathy and humor to create visually compelling illustrations of the stories I collected.

About This Book

This book uses illustrations to help you:

- Understand how absurd some real-life customer experiences have become
- Understand that these experiences occur across a wide range of industries including banking, healthcare, retail, cable, electricity, government, and more!
- Develop deeper empathy and caring for customers
- Understand the costs of customer experiences gone wrong are increasingly untenable to customers, employees, and companies
- Pinpoint some of the decisions that cause customer experiences to go wrong and spur you to make improvements

This is not a "methods" book. I do not believe there is a lack of methods, instead the problem is the lack of awareness of *how bad customer experience issues have gotten*. Along with the stories to raise your awareness of the issue, I offer exercises throughout the book to help you reflect on the concepts and stories and how they may relate to your organization. At the end of the book, I provide some suggestions to help you improve customer experiences. Their application depends on your organizational structure, challenges, and willingness to make changes.

Throughout the book, you will see gray boxes with these highlights:

Type of Box	Purpose
KEY POINT	Emphasizes an overall message in the chapter
THERE'S ALWAYS ANOTHER STORY	Provides an additional story to support the message in the chapter
BIG QUESTIONS	Exercises that help you think more deeply about stories or information presented, the decisions that may have been made, and the impact of those decisions **Note:** *The questions about your experience as a customer may seem contrary to the core User Experience principle "You are not your user," which emphasizes seeing things from the user's perspective, not your own. However, these questions are meant to highlight how common "customer experiences gone wrong" are.*

Who Should Read This Book?

The stories in the book are most relevant to people who work in organizations that have multiple products, departments/divisions, communication channels, etc., including:

- Product, Technology, and Business Leaders involved in product, service, technology, and budget decisions
- Designers, Researchers, Developers, and Product Managers who create products, services, and technology, and the teams that support them
- Employees of all types who work on initiatives or programs that provide products, services or technology

My hope is that understanding the details of these stories of customer experiences gone wrong strengthens your efforts to improve customer experiences within your organization.

> **KEY POINT:** This book is all about *"Understanding how customer experiences go wrong, so yours can go right!"*

Chapter 1:

IT'S EVERYWHERE

In April 2022, the tweet below appeared on my Twitter (now X) timeline and summed up what had been plaguing me and other customers for a while. Being a customer was frustrating because it seemed nothing went smoothly anymore. Simply put, "Nothing Works."[1]

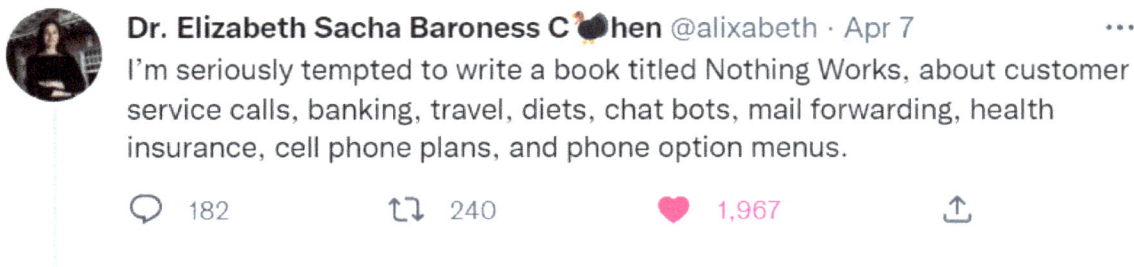

Dr. Elizabeth Sacha Baroness C🐔hen @alixabeth · Apr 7   ···

I'm seriously tempted to write a book titled Nothing Works, about customer service calls, banking, travel, diets, chat bots, mail forwarding, health insurance, cell phone plans, and phone option menus.

💬 182   ↻ 240   ♥ 1,967   ⬆

Figure 1.1 "Nothing Works" Tweet

Now, of course, this was somewhat exaggerated, but it captures how pervasive customer experience issues are, how they span across industries, how they impact different types of customer relationships, and how frustrating it feels to customers.

And I would add that decision makers don't realize how bad these experiences are nor would they want their company name associated with them.

What I Noticed...

About six or seven years ago, I began to notice that an item would stay on my to-do list for days or weeks at a time despite trying to get it accomplished. A payment error, a product that didn't work, an unexplained website glitch, inaccurate information, a disappearing order, etc. It was always something. And resolving the issue was never in my control. I was at the mercy of the company to get it resolved.

---

1. Dr. Elizabeth Sacha Baroness Cohen (@alixabeth), "I'm seriously tempted to write a book titled Nothing Works, about customer service calls, banking, travel, diets, chat bots, mail forwarding, health insurance, cell..." Twitter, April 7, 2022, 9:15 a.m., twitter.com/alixabeth/status/1512101710239567874.

It moved into "Nothing Works" territory when several issues began to overlap, along with the time-consuming and frustrating process of trying to resolve them. And often my money would be in limbo until the issue was resolved. Conversations with my family and friends helped me recognize that these types of customer experiences gone wrong were all too common, and there was an element of ridiculousness in how difficult these issues were to solve. There was a powerlessness because we alone could not get the issues resolved despite trying diligently. Everyone seemed to have one, two—or nine—stories that left a lasting negative impression.

THERE'S ALWAYS ANOTHER STORY

My personal "Nothing Works" moment. While my two stories here aren't the most serious, they show how simple transactions go wrong and no one at the business can explain what happened or why, but you have to do a lot of work to get them resolved. These overlapped in time as story #2 started the same day story #1 ends.

- Ordering Lunch: I had just inserted my credit card into the self-service kiosk when an error message informed me that the "order was terminated." I didn't know what that meant. But, my order didn't appear on the restaurant's order board. So, an employee rang up my order again and I paid. Two weeks later, I saw two charges on my credit card.
  - I had to call the restaurant twice and make a return trip to get a refund.

- Ordering Garden Supplies: I placed an order for store pickup at a nearby home improvement store. While the "Order Confirmation Email" had the correct pickup location, the "Order Ready Email" did not. The pickup location had been inexplicably changed to a store across town, which I didn't notice until I arrived at the store. No one could explain how my order location changed. They had to call the store where the order had been fulfilled and have them return the items so I could get a refund.
  - The original location no longer had all the items I needed in stock. So, I ended up going to a third store location to get the items.
  - I had to call twice because my refund wasn't processed the first time.
  - It took about 7 days to receive my refund.

What Are Customer Experiences Gone Wrong?

We'll start with a definition. **"Customer Experience"** is all the interactions a customer has with a company and can include pre-sales activities through purchase transactions and customer support to repeat business.

Let's be clear; the stories of customer experiences gone wrong in this book aren't about "the customer is always right" or "delighting customers" in these activities. It's more fundamental than that. It's about a **value exchange**. *Customer transactions or relationships are value exchanges where each party has to get something of value that they want from the other.*

For a customer, it may be:

- Completing a purchase, payment, return, etc.

- Getting what they paid for

- Receiving accurate information

- Getting a product, service or technology to work

- Getting errors or inaccurate information corrected

- Accessing a system

For a business, it may be:

- Money

- Repeat business

- Awareness

- Referrals

The customer expects a balanced transaction (a value exchange) as shown in Figure 1.2 where the customer is paying money for a phone:

Figure 1.2 Illustration of Value Exchange (Expectation)

"Unpaid Customer Labor"

Customers often do not have control over ultimately getting the value in the exchange, especially when there's an issue. For example, they can't physically get an item delivered, correct a payment error, process a refund or fix a system glitch. That's in the companies' control. Without the customer pushing for a resolution, they likely would not receive one. Their repeated efforts in time, money, and stress become *"Unpaid Customer Labor."*

"Unpaid Customer Labor" definition: The unwilling amount of time, money, and frustration repeatedly necessary from a customer to try to get a product, service or technology delivered or working, or resolve an issue related to it.

BIG QUESTIONS

- Can you think of a situation where you (as a customer) did not receive the value you expected/agreed upon in a transaction? How was the issue resolved?

- Can you think of a situation within your organization where a customer did not receive the value they expected/agreed upon? How was the issue resolved?

As you go through these stories, consider the value exchange.

- Who didn't receive the value that was initially agreed upon in the transaction?

- Why didn't they receive the agreed upon value?

- Who had control over the value that wasn't received?

- What business decisions were made to create this situation?

- What time, money, and stress were required by the customer while trying to get the initially agreed upon value?

Customer Experience Problems Are at an All-Time High—and Increasing

Beyond the stories in this book, the statistics show customer experience issues are increasing. According to the most recent edition of the National Customer Rage Survey conducted by Customer Care Measurement & Consulting (CCMC) in collaboration with The Center for Services Leadership (CSL) in the W. P. Carey School of Business at Arizona State University and released in January 2023,[2] most customers are having customer experience issues.

> **74% of customers have experienced a problem with a product or service in the past year. The number of customer problems has more than doubled since 1976.**

No type of industry is immune to customer experience problems. According to the survey:

- **Problems with small businesses (24%) are up by 50% (vs. 2003-2020)**

- **Problems with government agencies (12%) have tripled**

- **Problems with large businesses represent 62% of customers' most serious problems**

---

2. *Customer Care Measurement & Consulting (CCMC) in collaboration with The Center for Service Leadership (CSL) in the W. P. Carey School of Business at Arizona State University. The National Customer Rage Survey. A longitudinal survey of customer sentiments regarding complaint-handling in the United States. 10th ed., (Tempe: Arizona State University, January 2023), 19. Digital Copy.*

Everyday Life = The Most Common Customer Frustrations

Products and services used in everyday life lead the list of the most serious problems. Consider the top 10 Industries with the most serious customer problems include:[3]

- Computer/Internet
- Automobile (purchase, repair)
- Banking
- Cable/Satellite TV
- Airline
- Dining Out/Restaurants
- Gasoline sales
- Telephone (cell)
- Delivery services (FedEx, USPS, etc.)

Technology Is Everywhere

Technology has become integral to every aspect of daily life, including many customer transactions.

"Nearly three-quarters (72%) of customer interactions are now digital."[4]

As you will see in the stories in the book, technology plays a part in these customer experience problems as well.

When Technology Works, It's Great, When It Doesn't...

When technology works, it is fast, efficient, and effective, but it can also be problematic. A 2019 poll by Asurion showed more than 80% of Americans reported having

3. CCMC with CSL in the W. P. Carey School of Business at ASU. The National Customer Rage Survey. 19. Digital Copy.

4. MuleSoft Research in collaboration with Deloitte Digital. Digital Interactions—MuleSoft Research 2023 Connectivity Benchmark Report. Insights from over 1,050 IT leaders on the state of digital transformation. San Francisco: MuleSoft Research. February 2023. Online. https://hrinterests.com/wp- content/uploads/2023/ 11/2023-Connectivity-Benchmark-Report.pdf

a frustration with technology every day, and 53% experienced up to five tech frustrations a day. One in three (30%) get frustrated more than five times a day.[5]

As organizations champion digital transformations that include self-service options, the convenience to customers is expected to be wonderful. But it doesn't always work or work well. One downside of self-service is that an employee may not be present during the transaction to recognize and resolve issues. So, the onus is on the customer to initiate a resolution.

Hardened Against Human Interaction

At the same time, companies have increasingly implemented a technology-first approach to customer service or support hardening themselves against human interaction. And customers face a series of technology barriers when trying to reach a human who actually recognizes, understands, and can resolve the problem.

Technology's Inflexibility

There's an element of inflexibility when customers encounter an issue with technology; there's no reasoning with the spinning cursor on your computer or oblivious menu options on your phone. And don't even think about getting past the red error triangle. At best, the error message tells the customer what the error is and what to do next, but that's rare. Oftentimes a human has to initiate the resolution, but it is increasingly harder to contact a human within an organization.

Technology Becomes the Problem

And even in a case that doesn't start as a technology issue, it often becomes one when customers face layers of customer service or support technology that is ineffective and inflexible beyond a narrow set of issues it's designed to handle. An incomplete purchase, inaccurate data, a disappearing order, a disabled link or a nonsensical message—these aren't usually listed in the phone menu or as a chatbot option, even though you'll see in these stories that they happen all too often.

---

5. *Asurion in conjunction with OnePoll. "Top Tech Stresses That Frustrate Americans." Nashville: Asurion, 2019. Accessed January 13, 2024. https://www.asurion.com/press-releases/top-tech-stresses-that-frustrate americans/.*

What is the #1 frustration customers have when seeking help? Being forced to listen to long messages before being permitted to speak to a representative.[6]

One person I talked with summed up her feelings perfectly, "I hate phone trees."

Some would argue that technology has made such massive improvements that a few problems shouldn't be a big deal, but technology also can be ill-conceived, poorly executed, and inattentively monitored. Through the stories in this book and the statistics from the 2023 National Customer Rage Survey, you will see these problems aren't small in number and the impact on customers definitely isn't little.

The point is not to paint all technology negatively but to recognize where it falls short in customer experiences and improve it.

> **KEY POINT:** A problem that starts out small with a seemingly easy solution morphs into a bigger problem because of the inability to get the problem resolved quickly. The amount of lost time, financial peril, and emotional stress increases with each step.

Today is the Perfect Day to Improve Customer Experiences!

---

6. *CCMC with CSL in the W. P. Carey School of Business at ASU. The National Customer Rage Survey. 11. Digital Copy.*

Chapter 2:

COUNTING THE COSTS TO CUSTOMERS

The increase in the number of customer experience issues and the cost of customers' time, money, and stress is ridiculously high, and I would argue untenable.

According to the 2023 National Customer Rage Survey:

- **56% of customers felt the problem wasted their time (an average of 1–2 days of lost time)**

- **43% lost money (an average loss of $1,261)**

- **31% suffered emotional distress**

And most customers who complained report they had to put in a Very High to High Effort to complain—79%.[7]

As I began collecting these stories, a troubling pattern of "Unpaid Customer Labor" developed. Customers were spending outsized amounts of time, money, and emotional stress trying to get issues resolved. The value exchange is imbalanced.

Figure 2.1 Illustration of Value Exchange (Reality)

7. CCMC with CSL in the W. P. Carey School of Business at ASU. The National Customer Rage Survey. 11. Digital Copy.

THERE'S ALWAYS ANOTHER STORY

A customer was trying to order a set of cordless phones for her 104-year-old grand-mother who lives out of state. She looked online at three or four retailers but kept running into issues with pricing, availability, and confusing product descriptions.

She was finally able to place an order with one big-box retailer. Then, days later she received an email stating that they could not deliver the package to her grandmother's home because no one was available to sign for it. Not only was she unaware the order delivery required a signature, she also learned they had been attempting delivery at the wrong address in a different county.

Neither the package carrier nor the retailer could explain why this happened, but she was told they could reroute the package. Since no one would be available to sign for the package, a friend of the grandmother's tried to use the carrier's online tool to sign for the package so it could just be left at the door, but the online tool did not work. At an impasse, the customer canceled the order and waited for the refund to be processed.

Still needing to purchase the phones, she then ordered them online through a big-box electronics retailer where they could be picked up at a store near her grandmother's home to avoid the issue with the required signature. The order went through, but she didn't get a "Order Ready to Pick Up" email within 2 hours as promised. She checked online and then saw the pickup date had been changed to two weeks later. She called the store and was told the item was out of stock. She canceled the second order and waited for another refund.

She finally got the phones delivered by going into her local electronics store and having them ship the items with no signature required. Still frustrated by all the issues, the customer in this story had the following message for companies in Illustration 2.2:

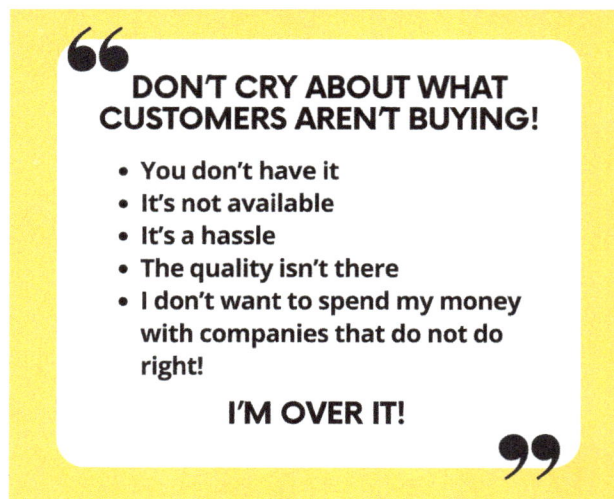

> ## " DON'T CRY ABOUT WHAT CUSTOMERS AREN'T BUYING!
>
> - **You don't have it**
> - **It's not available**
> - **It's a hassle**
> - **The quality isn't there**
> - **I don't want to spend my money with companies that do not do right!**
>
> ### I'M OVER IT! "

Figure 2.2 Customer Quotation

.**Figure 2.3** Illustration – Value Exchange (Expectation & Reality)

Notice in Figure 2.3 how the "value exchange" started as an equally agreed upon transaction, but the issues that came from trying to get the phones actually delivered made the transaction imbalanced—and not in the customer's favor. It was no longer the equally agreed upon transaction. And it was not in the customer's control to correct the impasse. She could only just start over several times.

Also, notice the Real Customer Receipt in the Reality image (Figure 2.3) shows how much the issue costs a customer beyond the price of the product or service they originally agreed upon. They never agreed to the transaction costing them additional time, more effort, or their emotional well-being.

Impact on Customers With the Least Resources

The impact of "Unpaid Customer Labor" on people who have few resources is also outsized. Even a small amount of money lost is significant. Living paycheck to paycheck, there's no room for lost money as it's for necessities like food, housing, childcare, transportation, etc.

Spare time to repeatedly call or follow up is also an issue for people with two jobs, in high demand or customer-facing roles as well as working parents to name a few groups. Many customers don't have money, time or emotional bandwidth to pay "Unpaid Customer Labor."

It's important to understand that the increasingly skewed value exchange and "Unpaid Customer Labor" affects quality of life. Lost time can't be recovered. Stress and frustration don't easily dissipate. Money is either lost or often out of the customer's control until the situation is resolved. There is powerlessness, anger, and exasperation in these situations. All of this can be infuriating and compound the stress of the original issue. As the customer in the telephone story says: "I can't do their job for them."

The impact of customer experience issues have multiplied morphing from having an issue occasionally to consumers having more than one issue to resolve at the time. And getting resolutions moving from hours or days to weeks and months and horribly enough sometimes years. This is a steep price to pay for anyone let alone someone with limited resources.

KEY POINT: While customers may not have OKRs (Objectives and Key Results) or run growth projections, like on the business side of the value exchange, they have their own calculus of when the cost of doing business with a company is untenable. They are calculating the Real Customer Receipt.

BIG QUESTIONS

- From the customer's perspective, compare the value exchange between the top "Expectation" and bottom "Reality" images in Figure 2.3. What are the differences?

- In your organization, consider the cost to a customer when a transaction goes wrong? How much would that customer's "Real Customer Receipt" total?

- In your organization, do you consider the impact of customer experience issues on customers with the least resources? How do you mitigate the harm to customers with the least resources?

- Can you think of a situation where you have had to pay "Unpaid Customer Labor"?

- What did it cost you in terms of time, money, and emotional stress? Did you decide to no longer do business with the company? At what point did you decide this?

- ***If you could draw a picture of that "Unpaid Customer Labor," what would it look like?***

KEY POINT: Business decisions that result in ill-conceived products, services or technology; unaddressed backlogs; poor monitoring; and ineffective customer service and support solutions have real financial costs to customers who will likely rethink which companies they will do business with.

"Unpaid Customer Labor" Illustration

Did the drawing you imagined look anything like the illustration on this page?

Figure 2.4 Illustration – "Unpaid Customer Labor"

KEY POINT: Companies' technology-first approach to customer service and support hardened them against human interaction similar to how prisons are hardened against unwanted entry. Customers struggle under the weight of repeated customer experience issues.

"Unpaid Customer Labor" Characteristics

Review the organizational characteristics identified that require "Unpaid Customer Labor" to get issues resolved. How many of these characteristics are present in your organization?

Figure 2.5 Illustration – "Unpaid Customer Labor' (annotated with numbers)

"Unpaid Customer Labor" Characteristics List

Review this list of organizational characteristics that require "Unpaid Customer Labor" to get issues resolved. How many of these characteristics are present in your organization?

1	**Limited Online Resolutions** *(Online problems can't be solved online)*
2	**Limited Customer Service Hours** *(Despite 24-hour sales availability, sales problems can only be dealt with during daytime business hours)*
3	**Inaccurate Product Information** *(Out of date, inconsistent, confusing, etc.)*
4	**Repeated Contact Required** *(Multiple efforts to get a problem acknowledged or resolved)*
5	**Financial Penalty** *(Delays, overcharges, penalties, etc.)*
6	**Long Waits** *(Hold times, resolutions, etc.)*
7	**Ineffective Customer Support** *(Limited help, multiple levels, varying quality)*
8	**Ineffective Feedback Loops** *(Customer or employee problems/feedback doesn't reach the people/teams who could provide resolutions)*
9	**Barriers to Contact** *(Missing or hard-to-find contact information)*
10	**Product Changes Disrupt Use** *(Frequent updates, redesigns, etc.)*
11	**System Issues** *(Unexplained errors, glitches, etc.)*
12	**Issue Unresolved** *(Despite notification or requesting help)*
13	**Transaction Data/Record Issue** *(No record of transaction, incorrect data, etc.)*
14	**Communications Issues** *(Confusing, unclear or lack of communication)*

Figure 2.6 List of organizational characteristics – "Unpaid Customer Labor"

Information Silos Are Real

Organizational characteristics that require "Unpaid Customer Labor" also show how companies operate in informational silos that negatively impact customers. Neither customers nor employees can get issues resolved easily and sometimes not at all. Whether it's an internal turf battle over which department "owns" what or employees laser-focusing on their department, product, team or journey, the lack of communication, coordination, feedback, and monitoring leads to the inability of customers (and employees) to get issues resolved.

Figure 2.7 Illustration – "Unpaid Customer Labor" – Peggy (highlighted)

The Power of Peggy

If customers happen to encounter a rare employee like "Peggy" who is featured in Figure 2.7 Illustration – "Unpaid Customer Labor" – Peggy (highlighted) above, their luck may improve.

"Peggy" represents that one long-term employee or small handful of long-term employees with the institutional knowledge to help customers untangle problems.

Peggy remembers old systems, discontinued products/services, former company names, grandfathered accounts, etc.

Sometimes, Peggy isn't so much a long-term employee but an employee who is especially diligent at becoming knowledgeable by asking questions, building relationships with other departments and keeping documentation. Most importantly, Peggy doesn't believe that everything the system shows is correct and doesn't dismiss issues. Peggy listens, asks questions, and researches to find the answer.

> **THERE'S ALWAYS ANOTHER STORY**
>
> A mom was looking at her account on her health insurance company's website and noticed medical information for one child was incorrectly listed under his brother's records. More unexplainable was that it wasn't a mix-up between her twins, who had the same birth date and similar names. It was a mix-up with her other child who had both a different birth date and last name. She called her health insurance company and spoke with three employees who insisted the information was correct and could not be changed.
>
> Their responses included:
>
> - Acting as though the mom was trying to do something fraudulent
> - Blaming the doctor (This did not make sense because the doctor had only seen one child and did not know the names or personal details of the other children)
> - Speaking rudely
>
> Concerned this would affect the treatment her son was receiving, the mother insisted on escalating the issue to get it resolved. Later that evening, she received a call from an employee who listened and was open to the possibility that the information in their system was incorrect. The employee was able to get the information corrected promptly.

Chapter 3:

COUNTING THE COSTS TO COMPANIES

While the COVID-19 pandemic has had a profound impact on businesses, particularly with labor shortages, it's also true that long-standing issues, including chronic understaffing, inadequate customer support systems, growing technical debt, increasingly complex products and services, and larger, siloed organizations have a lot to do with customer experience issues.

Some may argue that companies are making business decisions that increase shareholder value over things that would improve customer experiences. Whether designed to save money or push customers into more expensive products, I would argue that companies may not be counting all the cascading costs of poor customer experiences or are ready, willing, and able to pay the truly unpleasant costs.

As customer experience problems have increased, so have customers' efforts to resolve these problems. All this leaves customers increasingly angry, which has a domino effect on employees, operations, and company reputation, and that affects profitability. And the lack of awareness of how bad customer experiences have gotten contributes to more decisions that worsen customer experiences.

Revenue Loss

The cost of customer experiences going wrong is huge, with a total revenue loss of $887,860,440,931 for ineffective complaint handling.[8]

Some of the costs are obvious:

- Lost Customers
- Discounts and refunds to customers that have experienced issues
- Negative Word of Mouth
- Negative Media Coverage
- Marketing & Public Relations for reputation management

---

8. CCMC with CSL in the W. P. Carey School of Business at ASU. The National Customer Rage Survey. 14. Digital Copy.

- Marketing to replace lost customers (typically it costs more to acquire new customers)

- Marketing to existing customers who can't leave (monopoly, contract, convenience, etc.) but won't buy new products or services because of issues

- Executive Customer Relations

- Cost of Rework for Technology Issues

Legal and Regulatory

- Legal Fees

- Corrective Actions

- Fines

Other costs to be considered include:

- Lost productivity of multiple employees needed to resolve individual customer experience issues

- Increased security measures to protect employees and company property from angry customers

- Intense scrutiny from Congressional hearings, whistleblower lawsuits, etc.

Increased Employee Turnover

- Customer Service Center employees leave jobs because of burnout (the NICE WEM Global Survey reported a contact center attrition rate of 42% in 2021)[9]

- Employees in research, design, development, and other roles leave because of stymied efforts to improve technical debt and customer experiences

- Employees who leave because of poor reputation

Cascading Impact of Increased Employee Turnover

- Overtime pay for employees covering staffing shortages

- Dissatisfaction due to additional work because of staffing shortages

- Delayed projects because of staffing shortages/changes

- Employer Branding to recruit new candidates

- Initiatives to increase Employee Satisfaction and Retention

---

9. NICE. Contact Centers - from Attrition - to Retention 2022 NICE WEM GLOBAL SURVEY. This global survey report by NICE reveals the key drivers of attrition, the impact of the pandemic, and the best practices to improve employee engagement and retention in 2022. Ra'anana: Nice, 2022. Accessed January 13, 2023. https://www.nice.com/websites/prepared-agents-wem/files/2022-NICE-WFM-Global-Survey.pdf

- Increased customer complaints because of staffing shortages
- Work slowdowns, stoppages, and strikes due in part to staffing shortages

Customer Rage & Revenge

Increasingly customers are acting on their displeasure about their most serious problems. The 2023 National Customer Rage Survey shows:

- **63% of customers felt rage about their most serious problem. And they are becoming more vocal about it.**[10]
- **43% of customers yelled/raised their voice to express displeasure about their most serious problem (an increase from 35% in 2015).**[11]
- **And, 9% sought revenge for their most serious problem; that's tripled since 2020.**[12]

According to the most recent National Customer Rage Survey, complaining tended to increase customer rage. Those customers who expended higher effort to complain were 34% more likely to experience customer rage than those with lower effort.[13]

Employees Caught in the Middle

An unfortunate consequence of an increase in customer experiences gone wrong is the increasing number of customers seeking revenge because of product or service problems. Most times employees, on the receiving end of this displeasure, have very little to do with the business decisions that caused the problem. Or they don't have the information or access to correct the problem.

And employees are caught in between additional work to resolve the increase in complaints and dealing with the yelling, cursing, badgering as well as threats to take legal action or go to the media by customers seeking revenge.[14] None of this leads to job satisfaction or makes these companies desirable places to work.

---

10. CCMC with CSL in the W. P. Carey School of Business at ASU. The National Customer Rage Survey. 9. Digital Copy.

11. CCMC with CSL in the W. P. Carey School of Business at ASU. The National Customer Rage Survey. 9. Digital Copy.

12. CCMC with CSL in the W. P. Carey School of Business at ASU. The National Customer Rage Survey. 10. Digital Copy.

13. CCMC with CSL in the W. P. Carey School of Business at ASU. The National Customer Rage Survey. 11. Digital Copy.

14. CCMC with CSL in the W. P. Carey School of Business at ASU. The National Customer Rage Survey. 33. Digital Copy.

BIG QUESTIONS

- Do you believe companies are considering and calculating *all of these costs* when they make business decisions? Why not?

- What other costs are companies not considering?

- In your organization, what costs are being paid because of poor customer experiences?

- Are *all of these costs* recognized by decision makers/leaders in your organization?

- In your organization, has there been an increase in customer rage and revenge incidences?

- In your organization, what security measures have been deployed to protect employees and company property from angry customers?

The following chapters use illustrated stories to explore real-life customer experiences gone wrong to understand:

- What went wrong?

- Why did it go wrong?

- What decisions were involved?

- What are the costs to customers?

- How much "Unpaid Customer Labor" was required?

- How likely is the customer to continue a relationship with the company?

- What are the costs to employees?

- What are the costs to the company?

- What business decisions were made to create this situation?

- How could the company have eliminated or mitigated these issues?

Chapter 4:

PAYING AN ELECTRIC BILL

In this chapter, you'll have an opportunity to reflect on a customer's real-life experience and situation, and test your own understanding of what went wrong and the decisions that led to the issues. Here is some important background for the illustrated story below:

Customer: An elderly customer, concerned about slowing mail delivery, asks a family member to pay her electric bill online.

Company: Utility Company

- While the company, a monopoly regulated by the state, traditionally has had strong public support, it has faced increased scrutiny of its rates and operations from environmental groups, regulators, legislators, and the public in the last few years.

- The company markets additional products and services to existing customers through billing inserts, direct mail, etc.

Year(s): 2023-present

What Issues Did This Customer Experience?

Review the illustrated story below. Write down all the issues that the customer experienced.

Figure 4.1 Illustration – Paying an Electric Bill

Customer Experience Issues Identified

Compare the issues you wrote down with the list of issues identified below. How does your list compare?

Unclear/Confusing Communication	The email showed contrary information. It stated payment was canceled, returned, and accepted.
Unclear/Confusing Communication	Website alert shows $12.17 payment was returned but the customer did not make a $12.17 payment.
Transaction Data/Record Issue*	Representative #1 stated the confirmation number provided to the customer was incorrect and provided another confirmation number.
Ineffective Customer Support*	Representative #2 could not explain what happened to the payment and transferred the customer to another representative.
Ineffective Customer Support	Representative #3 could not explain what happened to the payment and needed to submit a billing department request to research the issue.
Transaction Data/Record Issue	Representatives stated the payment was returned for insufficient funds, but the customer had money in the account.
Long Resolution Period*	Customer was advised that the Billing Department can take up to one billing cycle (about 30 days) to complete its research.
Financial Penalty	Customer paid $12.17 fee because she was concerned about getting a disconnection notice.
Lack of Communication/ Notice*	Customer wasn't notified about the results of the research request (if it was ever completed).
Issue Unresolved*	With no follow-up provided on the Billing Department research, the customer will have to contact the company again (more "Unpaid Customer Labor").

* These issues were not included in the illustration for clarity but are listed here to show the full range of issues.

Figure 4.2 List of Issues – Paying an Electric Bill

BIG QUESTIONS

- Why did the customer receive an email that seemed to show the payment being returned, canceled, and accepted?

- Why couldn't the representatives explain with certainty what happened to the payment?

- Why were there two confirmation numbers?

- Did the customer receive what they were entitled to in the value exchange?

- After this experience, what would you do as a customer?

Counting the Costs

In trying to resolve the initial issue and receive a clear explanation of why the online payment was returned, the customer experienced several more issues that left her feeling distrustful of the utility company. Because the company is a monopoly regulated by the state and its customers don't have the option to change companies, it is immune from competitive pressure to improve its customer experiences. However, there is a domino effect of consequences that cost the company including those listed here:

Cost to Customer	Cost to Company
Lost Time • Customer made three calls to the utility company to try to resolve the issue • Customer made several calls to the bank to check on payment **Lost Money** • Customer paid $12.17 fee because of concern about a potential disconnection notice **Customer Distrust** • Customer's distrust of the company and its technology increased **Stress** • Customer became frustrated at the inability to receive a clear explanation • Customer became concerned about receiving a potential disconnection notice • Customer is frustrated that the issue remains unresolved	**Lost Productivity** • At least three employees were involved and spoke with customer • Billing Department research was potentially performed (unclear if it was performed) **Negative Word of Mouth** • Customer shared her experience with family members **Lost Revenue/Increased Costs** • Costs of direct mail and billing inserts to the now unreceptive customer • Increased costs of processing check payments as the customer no longer pays online

Figure 4.3 Customer & Company Costs – Paying an Electric Bill

THERE'S ALWAYS ANOTHER STORY

Another customer of this utility company had an additional electric line put into their home. About six months after installation, the customer had yet to get a bill for the additional service despite having called the utility twice about the issue.

KEY POINT: Companies are failing at explaining relatively simple issues where technology transactions are concerned. Customers want, expect, and need better communication and service.

Chapter 5:

REQUESTING A PRESCRIPTION REFILL

In this chapter as in the previous one, you'll have an opportunity to reflect on a customer's real-life experience and situation, and test your own understanding of what went wrong and the decisions that led to the issues. Here is some important background for the illustrated story below:

Customer: A working parent is looking to request a prescription refill from a primary care doctor using the telehealth service provided by her health insurance company. She has used the telehealth service previously for refills without an issue.

Industry: Health Insurance Company

Like many of its competitors, this health insurance company expanded its telehealth service during the COVID-19 pandemic and continued to market it as a benefit to customers.

Year: 2023-present

What Issues Did This Customer Experience?

Review the illustrated story below. Write down all the issues that the customer experienced.

Figure 5.1 Illustration – Requesting a Prescription Refill

April 2024 Update: Roughly six months after the initial issue, the customer's health insurance company still had not corrected the disabled telehealth link. Additionally, her doctor notified her that the health insurance company was closing its telehealth business within the next few months. The customer is frustrated because she relied on the convenience of the telehealth services for her family. This will affect her choice of health insurance companies during upcoming open enrollment.

Customer Experience Issues Identified

Compare the issues you wrote down with the list of issues identified below. How does your list compare?

System Issue	Customer encountered disabled telehealth link
Lack of Communication/Notice	Customers were not informed of the telehealth platform change.
Lack of On-screen Messaging	No website alert about the platform change.
Ineffective Customer Support	Representative #1 was rude* and could not provide accurate information on who to contact.
Ineffective Customer Support	Customer had to call three numbers to find out what the problem was.
Lack of Safeguards	Customer initially created a new account in the wrong system because there was no communication about a platform change.
Unnecessary/Duplicate Step*	Customer was required to create a new account because of the platform change.
Unnecessary/Duplicate Step*	Customer needed a medically unnecessary appointment with her doctor to get a refill because of the platform change.

These issues were not included in the illustration for clarity but are listed here to show the full range of issues.

Figure 5.2 List of Issues – Requesting a Prescription Refill

THERE'S ALWAYS ANOTHER STORY

On one project where I worked as a User Experience Researcher, the Product Manager informed me the company was going to disable some website features on a B2B product for a few months because it would be easier for a planned platform change. When I asked how many customers used one of the features, he seemed surprised by the question and didn't know the answer.

A few days later, he reported that about 25% of customers used that feature. While 25% wasn't huge, it was significant because some customers used the product daily in their work. I recommended research to better understand how customers were using the feature and mitigate the impact of a potential change. Unfortunately, I left the company soon after. So I don't know what happened with the project or if the Product Manager looked at customer usage when deciding on product changes going forward.

Counting the Costs

The decision to disable the old telehealth platform without notifying customers or providing a clear path of next steps cost this working parent and other customers in several ways. It also created extra work and inaccurate data for the company.

Cost to Customer	Cost to Company
Lost Time • Customer spent over one week trying to access the disabled links • Customer created an account in the incorrect system • Customer was required to create a new account • Customer was required to have a medically unnecessary appointment **Stress** • Customer was frustrated by links not working • Customer was frustrated at having to make several phone calls • Customer was frustrated the company did not tell customers about the platform change before disabling links • Customer was concerned medicine would run out • Customer was concerned personal information was now in the wrong system • Customer was annoyed at being spoken to rudely	**Lost Productivity** • Three workers were contacted regarding the disabled link **Inaccurate Data** • Customer created another account in the wrong system **Negative Word of Mouth** • Customer shared her experience with others **Customer Distrust** • Customer has become wary of new initiatives offered by the company • Customer considering changing health insurance companies during upcoming open enrollment

Figure 5.3 Customer & Company Costs – Requesting a Prescription Refill

THERE'S ALWAYS ANOTHER STORY

A few years ago, this customer wanted to get new glasses and planned to use her remaining Flexible Spending Account (FSA) money. In the last two weeks of the year, she scheduled an eye examination using the online system. When she arrived for the appointment, she was told that the location no longer performed eye exams.

The eye clinic employees did not understand how the customer could make an eye exam appointment since that location had stopped offering them. Or why an appointment notification wasn't sent to them when it was sent to the customer. Because of limited appointments during the holidays and her work schedule, she was not able to get an eye exam before the end of the year. The customer lost the several hundred dollars remaining in her FSA account. And the company lost a customer.

KEY POINT: Removing website features has consequences. It is important to:

- Understand how a change will impact customers and mitigate potential issues beforehand.

- Notify customers before system changes.

- Test to ensure all supporting systems are disabled/removed.

Chapter 6:

ORDERING A BANNER

In this chapter as in the others, you'll have an opportunity to reflect on a customer's real-life experience and situation, and test your own understanding of what went wrong and the decisions that led to the issues. Here is some important background for the illustrated story below:

Customer: Working mom with full-time job and small business needs a vinyl banner for an event in two days.

Industry: Drug Store Chain

Company regularly markets print products and services offered in their in-store photo labs.

Date: 2023

What Issues Did This Customer Experience?

Review the illustrated story below. Write down all the issues that the customer experienced.

Figure 6.1 Illustration – Ordering a Banner

Customer Experience Issues Identified

Compare the issues you wrote down with the list of issues identified below. How does your list compare?

System Issues	Customer had to upload a file nine times to get it submitted despite having the correct file size.
No On-Screen Messaging*	No additional information or troubleshooting messages about the file upload issue were available or appeared on screen.
Unnecessary/Duplicate Step*	Customer had to select a store location both before and after she uploaded the image. This occurred on the three orders she placed.
Lack of Communication/Notice	No one contacted the customer about the original order delay.
Issue Unresolved*	Store #1 Employee could not transfer the order to another store and could only cancel the order.
Ineffective Customer Support*	Customer Service Representative #1 did not have store phone numbers and asked the customer the number to another store before searching for it online.
Ineffective Customer Support	Store #2 Employee stated they could complete the order. Then, a few minutes later another Store #2 Employee stated they did not have the supplies.
Unnecessary/Duplicate Step	Customer had to place another order for the free paper banner because the vinyl banner was unavailable.
Ineffective Customer Support	Customer Service Representative #2 did not understand the issue or provide a resolution because of a language barrier.
Unclear/Confusing Communications*	Order Ready email and Order Pickup page on the website had conflicting information on when to notify the store of pickup.
Poor Product Quality	The banner was already peeling before its first use.
Issue Unresolved*	Three days after the original order, a Store #1 employee called to say the original order could not be completed. No employee had ever canceled the original order.

* These issues were not included in the illustration for clarity but are listed here to show the full range of issues.

Figure 6.2 List of Issues – Ordering a Banner

BIG QUESTIONS

- Why wasn't the supply inventory better tied to product availability?

- Why were vinyl supplies unavailable at two of the three stores?

- Why did Store #1 fail to notify customer?

- Why did Store #2 Employee incorrectly say they could print a vinyl order?

- Why did the Order Ready email and Order Pickup page on the website have conflicting information?

- Did the customer receive what they were entitled to in the value exchange?

- After this experience, what would you do as a customer?

- In your organization, how do you ensure information is consistent when different teams may "own" different parts of the journey? Ex. Email vs Website vs Store vs Customer Service

- Why did the Customer Service Representative not have accurate store information?

- In your organization, how do you ensure customer facing employees have relevant information to effectively assist customers?

Counting the Costs

The supply inventory and product availability/sales misalignment along with employee inaction and inaccurate information cost the customer and the company unnecessary time among other costs.

Cost to Customer	Cost to Company
Lost Time • Customer uploaded file nine times before it was accepted • Customer did not receive banner for two days • Customer made multiple calls • Customer had to make two additional orders • Customer had to make a second trip to pick up vinyl banner **Lost Money** • Banner had already begun to peel **Stress** • Customer was frustrated by repeated attempts to upload file • Customer was frustrated by repeated phone calls • Customer was annoyed by employee inaction and inaccurate information • Customer was frustrated by conflicting pickup information • Customer was worried about not having a banner for her event	**Lost Productivity** • At least six employees were involved between the stores and Customer Service **Lost Customer/Lost Revenue** • Free paper banner offered because of issues • Customer no longer uses store's photo lab services **Negative Word of Mouth** • Customer shared experience with others and at least one person did not use the store's photo lab services

Figure 6.3 Customer & Company Costs – Ordering a Banner

THERE'S ALWAYS ANOTHER STORY

A customer ordered several coats online from a children's retailer to be picked up at a nearby store. However, she never received an email saying the order was ready. Instead, she received an email saying the order had been put back because she hadn't picked it up.

When she called the 1-800 customer service number, she was told there was nothing they could do, but they would refund the payment to the gift cards she had used for the purchase.

She then called the store to see if the coats were still available. The store employee said they had just put the order back but could get it ready again. However when she arrived, the order was totally incorrect. It seems only half of the items were ever there and they were either the wrong size or color.

While she couldn't salvage the order, she ended up purchasing one coat from the store with her own money because the original order refund was still processing.

However, when she went to check the gift cards a month or so later, the balance was zero. When she called the retailer said they had no record of the gift card number or the transactions.

The customer hasn't had further time to escalate the issue to get her refund back, but she does not shop at the retailer anymore.

KEY POINT: Companies that rely on metrics alone would not capture the many failures at multiple touchpoints in this customer's experience. Qualitative data captured through customer interviews, observations, call center transcripts, etc. are important for understanding the details (and utter ridiculousness of what happens).

Chapter 7:

PICKING UP A PRESCRIPTION

In this chapter, you'll again have an opportunity to reflect on a customer's real-life experience and situation, and test your own understanding of what went wrong and the decisions that led to the issues. Here is some important background for the illustrated story below:

Customer: Working professional trying to find out if their prescription refill was ready from the pharmacy.

Industry: Grocery Store Pharmacy

A few times over the previous year, this pharmacy has posted signs notifying customers of staffing shortages and delays in prescriptions.

Year: 2022

What Issues Did This Customer Experience?

Review the illustrated story below. Write down all the issues that the customer experienced.

Figure 7.1 Illustration – Picking Up a Prescription

Customer Experience Issues Identified

Compare the issues you wrote down with the list of issues identified below. How does your list compare?

Transaction Data/Record Issue	The time the prescription would be ready changed four times.
Unclear/Confusing Communications	It was unclear if the system was referring to the previous day (ex. Thursday) or that day in the next week since the time the prescription would be ready only referred to the day of the week and not the date.
Limited Customer Service Hours*	Pharmacy had reduced hours over the past year but continued to add and market new services.
Lack of Communication/Notice*	There wasn't any message on the phone system about delays despite the prescription ready time changing four times.
Lack of Communication/Notice	Pharmacy Employee did not explain or apologize for the delay.

These issues were not included in the illustration for clarity but are listed here to show the full range of issues.

Figure 7.2 List of Issues – Picking Up a Prescription

48% of customers are getting nothing for complaining.[15]

BIG QUESTIONS
• Why didn't the prescription status system include the date as well as day of the week to prevent illogical or unreliable information?
• Why wasn't there a phone system message about the delays?
• Were staff, processes, and systems aligned to provide the level of service needed?
• What was the harm to customers and the company when prescriptions were delayed?
• Did the customer receive what they were entitled to in the value exchange?
• After this experience, what would you do as a customer?
• In your organization, have you encountered systems providing illogical or unreliable information?
• How do you prevent illogical or unreliable information from being provided?

15. CCMC with CSL in the W. P. Carey School of Business at ASU. The National Customer Rage Survey. 12. Digital Copy.

Counting the Costs

Staffing shortages, lack of communication, and unreliable prescription status information cost the company a pharmacy customer who then also made less frequent grocery purchases.

Cost to Customer	Cost to Company
Lost Time • Customer made repeated calls to check on prescription status • Customer was delayed in starting prescription (no health consequences) **Stress** • Customer was annoyed the prescription ready time kept changing • Customer was frustrated there was no message about the delays • Customer was frustrated by nonsensical information on prescription time • Customer was annoyed the long wait time was not acknowledged	**Lost Productivity** • Unreliable prescription status information increased the call volume for employees **Negative Word of Mouth** • Customer shared experience with others **Lost Customer/Lost Revenue** • Customer moved prescriptions to another pharmacy • Customer visited store less and made fewer purchases **Employee Stress** • Employees faced customers frustrated by delays • Employees had additional stress because of the delays

Figure 7.3 Customer & Company Costs – Picking Up a Prescription

---

THERE'S ALWAYS ANOTHER STORY

This customer had a previous issue with this pharmacy. She had received an automated phone message stating her prescription was ready. When she arrived at the store, only one of the requested prescriptions was ready. The other prescription wouldn't be ready for a couple of days requiring a second trip.

KEY POINT: Technology is often used to support operations and free up employees to do tasks that technology cannot or should not do. Ineffective or unreliable technology can exacerbate the impact of staffing shortages as more customers want to talk with an employee to ensure they receive accurate information. This creates additional work for employees working in places that were already short-staffed.

Chapter 8:

OPENING A BUSINESS CHECKING ACCOUNT

In this chapter as with the previous ones, you'll have an opportunity to reflect on a customer's real-life experience and situation, and test your own understanding of what went wrong and the decisions that led to the issues. Here is some important background for the illustrated story below:

Customer: Small business owner wanted to open a business checking account that specifically provided the ability to deposit checks.

Industry: Online Payment Platform & Bank Subsidiary

The online payment platform had sent the customer several marketing emails about opening a business checking account through its banking subsidiary.

Year: 2023

What Issues Did This Customer Experience?

Review the illustrated story below. Write down all the issues that the customer experienced.

Figure 8.1 Illustration – Opening a Business Checking Account

Customer Experience Issues Identified

Compare the issues you wrote down with the list of issues identified below. How does your list compare?

Missing or Inaccurate Product Information	The online account information listed depositing checks as a feature, and the directions for using the account app listed no limitations or exclusions about depositing checks based on previous business transactions.
Ineffective Customer Support	Representative provided the instructions for the check deposit feature until the customer provided a screenshot showing the feature wasn't there.
Unclear/Confusing Communications	Representative referred to "API" payments, a technical term the customer had never heard related to their account, as the reason the deposit check feature wasn't available.
Lack of Safeguards*	Customer supplied business and personal information to open an account that did not have the feature the online platform had advertised.

* This issue was not included in the illustration for clarity but is listed here to show the full range of issues.

Figure 8.2 List of Issues – Opening a Business Checking Account

---

BIG QUESTIONS

- What responsibility do the bank and the payment platform have for providing accurate product information?

- What risk(s) do the bank and the payment platform incur for inaccurate product information?

- Did the customer receive what they were entitled to in the value exchange?

- After this experience, what would you do as a customer?

- In your organization, how do you ensure that customer facing product information is accurate particularly with frequent product updates?

- In your organization, how do you ensure customer support teams have accurate product information particularly with frequent product updates?

Counting the Costs

After marketing to an existing customer, the payment platform lost a new banking customer and harmed its relationship with an existing customer because of inaccurate product information.

Cost to Customer	Cost to Company
Lost Time • Customer spent time opening account, trying to find the deposit feature, speaking with Customer Service, and closing the account **Privacy Concerns** • Customer shared personal and business information to open the account **Stress** • Customer became angry at wasted time and inaccurate information • Customer was annoyed because the representative's API payments explanation didn't make sense **Distrust** • Customer does not trust the online payment platform	**Lost Productivity** • Representative spent time providing inaccurate instructions before finding out why the feature wasn't available **Negative Word of Mouth** • Customer shared experience with others **Lost Customer/Lost Revenue** • Customer closed checking account the day after opening it • Customer unsubscribed to marketing emails • Customer is looking for a new payment processor **Employee Stress** • Employee had to respond to angry customer

Figure 8.3 Customer & Company Costs – Opening a Business Checking Account

- Selling Digital Products: This customer encountered a similar problem when the same company advertised the ability to sell digital products through its platform. Only it did not work seamlessly.

  Once a consumer ordered a digital product, they should have received an email confirming the order with a link to download the digital product. Unfortunately, that email generated through the platform listed a mailing address that was unnecessary for a digital product. The mailing address couldn't be removed from the email which could be confusing to consumers. When the business owner reached out for help, the company representative provided incorrect instructions, stating the mailing address could be edited or removed. The business owner had to explain that part of the email could not be edited.

  The representative told the business owner they still could send digital product buyers the email with both the incorrect mailing address and a link to the digital product. This was neither a user-friendly nor professional solution. The customer decided to use another platform to sell digital products.

- Opening a Business Checking Account (Different Bank) - After seeing several online ads from a large bank promoting a cash bonus for opening a business checking account, the customer completed the online application. They received an email stating they had successfully esigned their documents along with an account signature card with the new account number.

  However, instead receiving required disclosure documents, the customer received an error message with instructions to contact the bank. She contacted the bank and spoke with a VP who said they did not see the account in their system, but would check the next day (a Saturday) to see if it appeared. When the account did not appear on Monday, the VP recommended the customer stop by the closest branch to open the account instead. The customer eventually went into the branch and completed the whole process again with a bank representative.

KEY POINT(S): Companies often prioritize marketing and sales efforts over resolving technical debt or keeping product information updated. While marketing products and services to existing customers is a smart strategy, maintaining accurate product information is equally as important in order to retain customers.

Additionally, Companies often think they have a marketing or sales problem when they actually have a user experience problem. They are unaware that customers are not able to use their products or services with ease.

Chapter 9:

UPGRADING CABLE CHANNELS

In this chapter as with the previous ones, you'll have an opportunity to reflect on a customer's real-life experience and situation, and test your own understanding of what went wrong and the decisions that led to the issues. Here is some important background for the illustrated story below:

Customer: Customer called to request separate billing for their internet service. The customer is scheduled to go on a business trip at the end of the week and is awaiting a call about medical test results for a relative who was recently in the hospital.

Industry: Cable Company

- During the call, the representative offered the customer a special deal that would add more cable channels for a lower rate with no other account changes.

- The Cable Industry historically has had low customer satisfaction rates.[16]

Year: 2022

---

16. *The American Customer Satisfaction Index (ACSI). "Subscription TV Service - The American Customer Satisfaction Index." Theacsi.org: The American Customer Satisfaction Index (ACSI), 2023. Accessed January 15, 2024. https://theacsi.org/industries/telecommunications-and-information/subscription-tv-service/.*

What Issues Did This Customer Experience?

Review the illustrated story below. Write down all the issues that the customer experienced.

Figure 9.1 Illustration – Upgrading Cable Channels

Customer Experience Issues Identified

Compare the issues you wrote down with the list of issues identified below. How does your list compare?

System Issue	Customer could not submit an order form.
Transaction Data/Record Issue	Representative #2 could not locate a copy of the original order form.
Lack of Safeguards	Customer's phone number was given away without the customer's permission.
Lack of Communication/Notice	The customer received no communication that the phone number was changing.
Limited Customer Service Hours/ Options*	The office that handled the customer's order closed at 6pm.
Ineffective Customer Support	Customer spent several hours on calls, email and chat messages trying to find out what happened to her phone number.
Transaction Data/Record Issue	Executive Customer Support Representative initially stated the customer had a contract.
Unclear/Confusing Information*	Executive Customer Support Representative used legalese to explain the situation.

These issues were not included in the illustration for clarity but are listed here to show the full range of issues.

Figure 9.2 List of Issues – Upgrading Cable Channels

The 2023 National Customer Rage Survey shows:

> **Customers want explanations and answers in understandable language.**
>
> - **25% want "An explanation of why the problem occurred"**
>
> - **14% want "An answer in everyday language/more than scripted response to your problem" (which is one of the least likely things customers get)[17]**

---

17. *CCMC with CSL in the W. P. Carey School of Business at ASU. The National Customer Rage Survey. 35. Digital Copy.*

Counting the Costs

After calling simply to request separate internet billing, the customer was offered and accepted a deal on expanding cable channels. The result was a four-day service disruption, postponed trip with hotel cancellation fee, and other hassles with many company employees involved including Executive Customer Support.

Cost to Customer	Cost to Company
Lost Time • Customer had repeated calls, emails, and chat messages to find out what happened to the phone service and get the issue resolved • Customer had to search for new hotel and flight arrangements • It took four days for the phone number to be restored **Lost Money** • Customer paid hotel cancellation fee (cable company eventually provided a bill credit for most of the fee) **Stress** • Customer became angry their phone number was given away • Customer worried over missing a doctor's call • Customer worried over delaying trip • Customer stressed at having to find new hotel and flight reservations	**Lost Productivity** • Customer was in contact with at least eight employees while trying to get the issue resolved **Negative Word of Mouth** • Customer shared experience with others **Bill Credit** • Company provided bill credit for most of the hotel cancellation fee **Lost Revenue** • While customer keeps their service with the company, they are not receptive to any marketing offers and would readily leave for a better deal **Employee Stress** • Employees had to respond to angry customer

Figure 9.3 Customer & Company Costs – Upgrading Cable Channels

KEY POINT: The company never explained the specific cause of the error other than to say it was a "system issue." It also lacked important safeguards, including retaining a copy of the order form and requiring customer permission to give away a phone number.

Chapter 10:

APPLYING FOR UNEMPLOYMENT BENEFITS

This chapter is different in that it involves my experience with the Virginia Employment Commission (VEC), a state government agency. As you reflect on this real-life experience and situation, you will see a similar but amplified pattern of customer service failures and technology barriers as in the other stories. Test your own understanding of what went wrong and the decisions that led to the issues. Here is some important background for the illustrated story below:

Claimant*: During the pandemic, a working professional experienced a layoff and tried to apply for unemployment benefits.

Industry: State Employment Commission (Government)

- Like many state agencies, the VEC faced an unprecedented number of unemployment claims during the COVID-19 pandemic.

- However, the VEC, which was already underfunded, understaffed, and in need of a long-delayed technology upgrade, performed poorly.

- Three years, a technology upgrade, changes in administrations/leadership, it continues to perform poorly.[18]

Year(s): 2021-present

* The term claimant is used in this story instead of customer to be consistent with how the VEC refers to people filing unemployment claims.

---

18. Rankin, Sarah, and Bill Fitzgerald. "When Virginia Employment Commission expects to wipe out pandemic unemployment backlog." WTVR.com, December 7, 2023. Accessed 17 January 17, 2024. https://www.wtvr.com/news/local-news/when-virginia-employment-commission-expects-to-wipe-out-pande mic-unemployment-backlog-dec-7-2023.

What Issues Did This Customer Experience?

Review the three-page illustrated story that starts below. Write down all the issues that the customer experienced.

Figure 10.1 Illustration – Applying for Unemployment Benefits (May 2021–Oct 2022)

I finally got a letter saying that I won my appeal and that my benefits starting date would be changed.

The letter did not provide any instructions on how to get my benefits and neither did the VEC website.

Over the next several months, I called and received different responses on how to get the benefits I was entitled to from the appeal.

After months of waiting I wrote an OP-ED titled "HOW LONG IS TO LONG FOR THE VEC?" published on the Virginia Mercury news site.

The next day I received an email from the VEC with instructions to visit their online portal for new documents.

Unfortunately, I was locked out of the portal. I finally spoke with a representative who mailed me a series of forms to fill out.

She gave me specific instructions on how to note on the form that the Temporary Waiver for Job Search Requirements was in effect for two weeks I was due benefits.

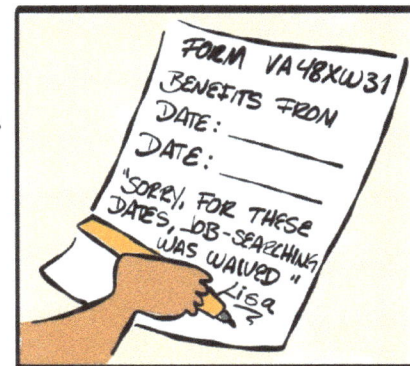

I received two documents in the online portal requesting information again on my work search.

Again, I filled out the forms noting that the Temporary Waiver was in effect.

Figure 10.2 Illustration – Applying for Unemployment Benefits (Dec 2022–Aug 2023)

Then, another set came in the mail all asking about job search information that I had already provided.

I finally gave up and sent the forms with the same information I had provided already.

It's been over 900 days since this started. To this day (December 18, 2023), I still don't have all the benefits I am entitled too.

And over 41000 families today are still waiting for their appeals. Meanwhile rent, groceries, car payments, healthcare, childcare, and more are still due.

HOW LONG IS TOO LONG?

Instead of getting the benefits that we are entitled to and that means for us childcare, mortgage, medicines...

...what we get are technology barriers, silence, bureaucracy, and a system that doesn't work.

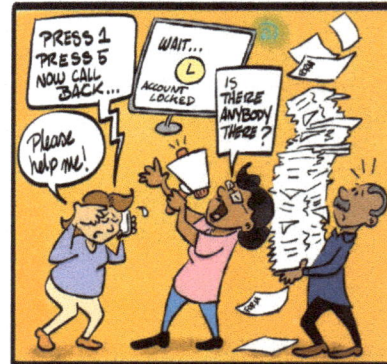

Figure 10.3 Illustration – Applying for Unemployment Benefits (Sept 2023–Dec 2023)

JUNE 2024 UPDATE: When I did not receive any communication from the VEC after I submitted the duplicate information in September 2023 that they requested, I called the main Customer Service Number. I could not reach a person, but I did receive an automated report on my claim. While the report recited a list of Issue Numbers with the status "Pending Under Review," the issues themselves were not identified by a name nor were any dates provided. This did not provide me with any information that would help resolve the issue. It is now three years since the issue started and 18 months since I won my appeal, yet I have not received the remaining benefits that I'm due.

Customer Experience Issues Identified

Compare the issues you wrote down with the list of issues identified below. How does your list compare?

Lack of Communication/Notice	Website information instructed the claimant to apply online, but a PIN that was needed was not available online.
System Issue	Automated phone system hung upon the claimant numerous times.
Ineffective Customer Support	Besides making repeated calls, including calling 26 times over one month to try to file a claim, the claimant also experienced long call wait times.*
Lack of Communication/Notice	Claimant did not receive any information about the appeal for over a year.
Lack of Communication/Notice	Appeal Determination Letter notified the claimant that they had won their appeal but provided no next steps on how to collect benefits.
Lack of Communication/Notice*	No information was provided on the VEC website or online portal on what the next steps were for receiving benefits after winning an appeal.
Ineffective Customer Support*	No ability for the claimant to communicate the issue through the online portal.
System Issue*	Automated phone system instructed the claimant to call the same number they had just called.
System Issue*	Claimant became locked out of the online portal on two occasions.

Lack of Communication/Notice*	Claimant has not received any response to the email complaint submitted four months before.
Unnecessary/Duplicate Step	Claimant completed three sets of forms requiring the same information without explanation.
Unclear/Confusing Communication	Forms received by the claimant asked for job search information for the two weeks when the Temporary Waiver of Job Search Requirement was still in effect.
Financial Penalty	Claimant has waited for over 3 years for the missing benefits.
Issue Unresolved*	Over 1.5 years since winning the appeal and over 3 years since filing the appeal, the claimant has not received the remaining benefits they were entitled to.
Ineffective Customer Support*	Phone number listed on the website for complaints that could not be handled by the main Customer Service number rang to the main Customer Service number.

These issues were not included in the illustrations for clarity but are listed here to show the full range of issues.

Figure 10.4 List of Issues – Applying for Unemployment Benefits

---

BIG QUESTIONS

- Why has the ineffective communication persisted?
- What expectations of customer service should be expected of government services?
- Since government agencies do not have competitive pressure to improve services, what influences them to improve?
- What has been the impact on the claimants that have gone without benefits?
- Is there a value exchange? What is it?
- After this experience, what would you do as a claimant?

Counting the Costs

Despite following all outlined instructions to apply for and receive unemployment benefits, the claimant and tens of thousands of others have encountered obstacle after obstacle over several years, and have yet to see any benefits.

Cost to Customer	Cost to VEC
Lost Time • Claimant has made numerous phone calls and experienced long wait times, hang-ups, and receiving incorrect information • Claimant filled out duplicate forms **Lost Money** • Claimant entitled to benefits from over 3 years ago **Stress** • Claimant experienced frustration at the long delays • Claimant has been frustrated at the lack of communication • Claimant frustrated by the lack of checks and balances to influence the VEC to improve	**Lost Productivity** • Claimant spoke with at least 12 employees over the three-year period. • Persistent lack of communication spurned increased calls by claimants trying to get answers **Public Distrust** • Persistent negative media coverage • VEC faced lawsuit • Increased scrutiny from lawmakers **Negative Word of Mouth** • Claimant shared experiences with other and wrote an op-ed **Employee Stress** • Staff encounter claimants who are angry at long delays • Persistent staff shortages and a layoff exacerbated workload pressure on staff

Figure 10.5 Customer & Company Costs – Applying for Unemployment Benefits

KEY POINT: While the government has different mandates and constraints than private industry, a value exchange is still involved. Government provides public services funded by taxpayers, and people follow the rules required to receive benefits. The long wait times, ineffective communications, technology barriers, and delay in benefits continue to harm not just claimants and their families, but also providers of the products and services they still need to pay, like for childcare, rent, medicine, car payments, etc.

As was concluded in McKinsey's 2022 State of States Survey, a benchmark study of 12 state government services, "Our research is clear. Customer experience matters, with higher satisfaction levels driving greater trust in government, cost savings, and fewer negative media mentions."[19]

19. Ashka Dave et al. "OMV to Medicaid: Improving customer experience in state services." New York: McKinsey, 2022. Accessed January 17, 2024. https://www.mckinsey.com/industries/public-sector/our-insights/governments-can-deliver exceptional-customer-experiences-heres-how.

Chapter 11:

WHAT'S NEXT?

After reading these real-life stories, I think you'll agree these customer experiences went ridiculously wrong and the costs to customers, employees, and companies are untenable for lasting relationships. And as the National Customer Rage Survey data shows, these aren't isolated incidents. Organizations need to improve customer experiences now.

This includes examining and resolving where technology exacerbates customer experience issues. Technology is beneficial, but it must be well-conceived, executed with care, and monitored appropriately to ensure "Unpaid Customer Labor" isn't a byproduct.

Here are a few suggestions as a starting point.

- Devote resources to qualitative research methods including customer interviews, observations, usability testing, and user experience audits along with surveys and analytics to begin understanding customers' wants and needs and how your product, service or technology is or is not working.

- Proactively seek out where customer experiences go wrong in your organization. This is vital to improvement. This can include mining call center recordings, chat messages, complaint forms, search logs, and other customer contact channels to better understand issues.

- Use Figure 2.6 List of Organizational Characteristics to identify and mitigate ways your organization requires "Unpaid Customer Labor."

- Calculate the cost to employees and the company for poor customer experiences . Decide if the company is ready, willing, and able to pay them.

- Educate all employees including leaders on the long tail of cascading costs of poor customer experiences. Use the Revenue Loss section on Page 25-27 as a reference point.

- Promote robust customer and employee feedback loops to connect issues with the teams responsible for and capable of improving them.

    o Identify your informational silos and work to break them down.

    o Seek ways to measure how well information is shared throughout the organization.

- Promote not just "ownership" of particular products, stages of the journey, processes, etc. but also accountability for information-sharing and the customer experience.

- Find the "Peggy or Peggies" in your organization to better understand knowledge gaps, the informational silos, technology issues, etc. that affect customer experiences.

- Ensure that Peggy receives recognition and compensation for the value her assistance brings to improving customer experiences for the organization.

- Incentivize improving customer experiences and reducing technical debt.

- Create Customer Journey Maps to help evaluate the multi-touchpoint experiences many customers encounter. Map touchpoints as well as emotions involved so you can identify pain points and opportunities for improvement. Add the costs of poor experiences to maps.

- Utilize Service Design techniques to expand your Customer Journey Map into a Service Blueprint by adding information about the people, processes, and systems that support the customer journey on the map. This will help you identify the disconnect between what the customer is experiencing (the frontstage) and internal parts of the organization (backstage).

These methods will help improve customer experiences. Over time, you will naturally discover new methods, systems, and processes that will keep you on this all too important path. Along the way, an important North Star to remember is that customers have wants and needs, and meeting these should be as important as meeting business goals. It is, after all, a value exchange.

Today is the Perfect Day to Improve Customer Experiences!

Do You Have a Story of Customer Experience Gone Wrong or "Unpaid Customer Labor"?

Whew! These stories are a lot, but they aren't the only ones. Do you have a story of a customer experience gone wrong that you would like to share? I'm collecting stories and would love to hear yours. If you have a story of "Unpaid Customer Labor" to share, please fill out this form: **https://tinyurl.com/UnpaidCustomerLaborStory**

INDEX

ABOUT THE AUTHOR

Lisa D. Dance

UX Consultant & Founder

*Service*Ease

Lisa D. Dance's path into the user experience field started when she was 9 years old and told her mother she wouldn't shop at a particular store anymore because of how they treated their customers. Ever since then she's studied customer interactions and strategized ways to improve them for both customers and businesses.

Today, Lisa D. Dance helps organizations create online and offline experiences that don't frustrate or harm people. As a User Experience Research Consultant/Founder at ServiceEase, her work spans UX Research and Strategy to Interactive Prototyping and Usability Testing. Lisa has worked with enterprise organizations like Indeed and Genworth Financial, whose website won 17 national and international awards over a two-year period.

Lisa earned a Bachelor of Arts in Political Science, Post-Baccalaureate Certificate in Marketing, and an Interaction Design Certificate. She is a contributor to CMSWire and a frequent public speaker and workshop facilitator on topics related to User Experience, Customer Experience, Ethical Research & Design, and Technology's Impact on People. In 2019, she created the "3Q Do No Harm Framework" to help teams identify and mitigate potential harms of technology before launch.

Lisa's website is ServiceEase.net.

ABOUT THE ILLUSTRATOR

Antonio Meza

Do you have big ideas and you want to get buy-in from potential users, investors or just people with the power to green light your projects? Better call Antonio Meza.

Antonio is a business consultant with visual superpowers. His passion is to listen to your ideas and help you communicate them in a clear and compelling way so you can obtain the buy-in you need to make them happen.

He was born in Mexico where he got his roots, and has been living in France for the last 18 years where he got his wings. He has worked in the audiovisual industry and also in the fields of soft skills training and coaching.

Because of his unique capacity to listen, understand and represent ideas he has travelled around the world providing graphic facilitation and storytelling services. He has illustrated 16 books in the fields of business and personal development. He is based in Paris, France where he produces animated videos, storyboards and illustrations to support purposeful projects and ventures.

Antonio's website is antoons.net.